IT'S WRITTEN
in the
STARS

Written by

Christina Rosso-Schneider

T0364070

RP Minis®
Hachette Book Group
1290 Avenue of the Americas, New York, NY 10104
www.runningpress.com
@Running_Press

First Edition: August 2023

Published by RP Minis, an imprint of Perseus Books, LLC, a subsidiary of Hachette Book Group, Inc. The RP Minis name and logo is a trademark of the Hachette Book Group.

Running Press books may be purchased in bulk for business, educational, or promotional use. For more information, please contact your local bookseller or the Hachette Book Group Special Markets Department at Special.Markets@hbgusa.com.

Hello, stargazers! Inside this mini kit you'll find eight wooden magnets of popular constellations and a glow-in-the-dark double-sided poster of constellations in the Northern and Southern Hemispheres. We've also included little profiles about each constellation featured on the magnets, with fun facts and mythology.

Before we jump in, let's start with some basics. Constellations are famous around the world for several reasons. Ancient civilizations used the stars to figure out when to plant and harvest their crops. Over time, they named these clusters of stars, creating the first constellations, by weaving in their culture's mythologies of creation and magic. Stars also aided sailors in their travels and were used to gaze into the future. Many kings employed astronomers to chart the stars and tell them prophecies.

In this mini book, when we talk about constellations in the Northern Hemisphere, we're referring to the half of Earth that is north of the equator. So, constellations that appear in the Southern Hemisphere are typically seen only south of the equator. That said, some constellations are visible in both hemispheres because they shift seasonally. Now let's get into some star stories.

URSA MAJOR
& URSA MINOR

Identifying Features: Ursa Major is the most universally recognized constellation because it's always visible in the Northern Hemisphere. With seven stars forming part of the back of the constellation, the Big Dipper also helps stargazers find the North Star, or Polaris. The Big Dipper is made of two parts, a bowl and a handle. Make a line with the two stars of the Big Dipper's outer edge, Dubhe and Merak (known as the Pointers), extend that line up into the sky, and you'll find the North Star.

Ursa Minor is best distinguished from Ursa Major by its size and contains Polaris, the North Star.

Hemisphere: Northern. Both the Ursa Major and Ursa Minor are visible year-round in this hemisphere. The Ursa Major can be found at latitudes between +90° and -30° and the Ursa Minor at latitudes between between +90° and -10°. Ursa Major can be seen in the Southern Hemisphere April through June.

Fun Fact: The Big and Little Dippers are known as asterisms, well-known, prominent patterns of stars that are smaller or sometimes part of a larger constellation.

Ursa Minor is a circumpolar constellation, which means it never sets in the northern sky and is always above the horizon.

Nearby Constellations: Bootes, Draco, Leo, Camelopardalis, and Cepheus.

Mythos: In Greek mythology, Hera discovered Zeus was having an affair with a beautiful mortal

named Callisto and turned her into a bear. Zeus put Callisto in the sky along with her son, Arcas, who became the Little Bear (the Little Dipper).

Like many stars, stories of the Ursa Major are famous around the world and not just limited to Greek and Roman mythology. In China, for instance, the Big Dipper is known as Běidǒu

(北斗). It is seen as the throne of Shàngdì (上帝), the Supreme Deity or Emperor in ancient Chinese religion from the Shang Dynasty to later Daoism. In some stories, the Big Dipper is the chariot of the Emperor controlling the sky. Chinese astronomy also has the Little Dipper, which is known as the South Dipper Six Stars (南斗六星). It is in Sagittarius. In a Chinese proverb, the Little Dipper is believed to mark life, while the Big Dipper marks death.

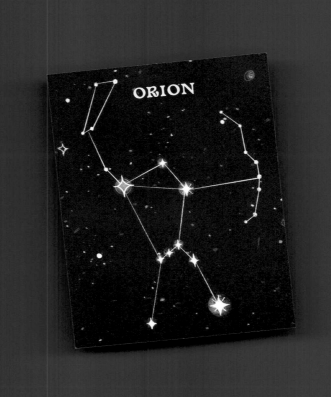

Identifying Features: One of the largest and most recognizable constellations, Orion is often said to look like a person wearing a belt and holding weapons.

Hemisphere: Northern and is best seen from November through February between latitudes +85° and -75° but can be seen in the Southern Hemisphere in the summer months. (The hunter will appear upside down!)

Fun Fact: Orion's Belt is made up of three stars: Mintaka, Alnilam, and Alnitak.

Nearby Constellations: Eridanus, Gemini, Lepus, Monoceros, and Taurus.

Mythos: In Greek mythology, Orion was the son of Poseidon and a gifted hunter. He regularly hunted with Artemis, the goddess of the hunt, and, depending on the myth, was killed either by the goddess's bow or by the sting of the great scorpion, who later became the constellation Scorpius. Orion also has connections to the Babylonian hero Gilgamesh.

TAURUS

Identifying Features: The red star, Aldebaran, as well as a star cluster known as the Pleiades, can be found in the Taurus constellation.

Hemisphere: Northern. It's best seen in autumn and winter at latitudes between +90° and -65°, but can be seen in the Southern Hemisphere from late spring through summer.

Fun Fact: Taurus is the second sign in the Zodiac and represents those born between April 20 to May 20. It connects to the Hierophant, the fifth card in the Major Arcana, which symbolizes tradition, structures, and values.

Nearby Constellations: Orion, Ceres, Auriga, Aries, Gemini, Eridanus, and Perseus.

Mythos: As one of the oldest recognized constellations, the mythos surrounding this constellation goes back to the Bronze Age spanning from the Greeks to the Babylonians. In one Greek myth, Hercules must capture the Cretan Bull as his seventh of twelve labors.

In Babylonian mythology, the bull was known as Gugalanna, the Bull of Heaven. Gugalanna appears in the epic of Gilgamesh. After Gilgamesh denies the goddess Ishtar's advances, she asks her father, Anu, to send the Bull of Heaven to kill Gilgamesh, even though it will cause famine.

However, Gilgamesh and his friend Enkidu manage to kill the creature. Gilgamesh is represented by the constellation Orion, which is next to Taurus. Some say the two are still fighting their battle in the skies to this day.

Identifying features: Leo is easy to identify, with the bright stars Regulus marking the lion's heart and Denebola its tail, and an asterism known as the Sickle shaping the lion's mane and shoulders. The Sickle looks like a backward question mark and is most visible in the spring.

Hemisphere: Northern. It's best seen January to June at latitudes between +90° and -65°. It can be seen in the Southern Hemisphere in the summer and fall months.

Fun Fact: Leo is the fifth sign in the Zodiac and represents those born between July 22 to August

22. It also has connections to tarot, specifically the eighth card in the Major Arcana, Strength, where in the Rider Waite Smith deck a maiden is depicted petting a lion. This card symbolizes quiet strength, courage, and perseverance.

Nearby Constellations: Cancer, Crater, Hydra, Lynx, Ursa Major, and Virgo.

Mythos: In Latin, Leo translates to lion. Like Taurus, Leo has a connection to Hercules and is often said to represent the Nemean Lion. This was a monster who was impervious to iron, bronze, and stone that Hercules slayed as his first labor.

Identifying Features: Gemini is located northeast of Orion and between Taurus and Cancer. The two brightest stars in the constellation represent the heads of the twins Castor and Pollux, while fainter stars outline the twin's bodies.

Hemisphere: Northern. It's best seen in the winter and spring months at latitudes between +90° and -60° but can be seen in the Southern Hemisphere in the summer.

Fun Fact: Gemini is the third sign in the Zodiac and represents those born between

May 21 and June 21. It also connects to the Lovers tarot card, the sixth in the Major Arcana, which symbolizes integration, partnership, and communication.

Nearby Constellations: Auriga, Cancer, Canis Minor, Lynx, Orion, and Taurus.

Mythos: In cultures around the world, the Gemini constellation represents twins. In Egyptian astrology, the constellation formed twin goats, and in Arabian astrology, twin peacocks. In Greek mythology, Gemini represents the twins Castor and Pollux. Their mother was Leda, a woman so beautiful that Zeus turned into a swan to seduce her. Their coupling led to the birth of

Pollux and Castor's father, who was the King of Sparta. The tale differs depending on the culture, but the ending is always the same. When Castor, the mortal twin, dies in battle, and Pollux begged Zeus to grant his brother immortality, so the god of thunder agreed, placing the twins in the sky as a constellation, so they could be together forever.

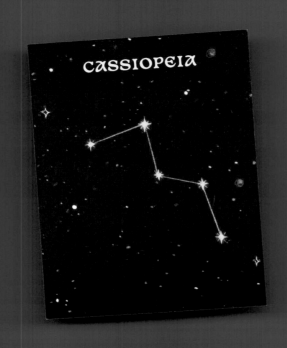

Identifying Features: Famous for its distinctive W shape, which is an asterism formed by five bright stars in the constellation: Segin, Ruchbah, Gamma Cassiopeiae, Schedar, and Caph. She looks like she is sitting on a throne combing her hair.

Hemisphere: Northern. It's visible year-round at latitudes between +90° and -20°, but can be seen in the Southern Hemisphere in the last spring.

Fun Fact: Cassiopeia is the twenty-fifth largest constellation in the sky.

Nearby Constellations: Andromeda, Camelopardalis, Cepheus, Lacerta, and Perseus.

Mythos: Cassiopeia, the Queen of Ethiopia, was known for her vanity and once boasted she was more beautiful than the Nereids, the fifty sea nymphs fathered by the Titan Nereus. Enraged, the nymphs begged Poseidon to punish the queen, so the god sent Cetus, a sea monster. The oracle told Cassiopeia and her husband, King Cepheus, they had to sacrifice their daughter, Andromeda, to the monster to appease Poseidon. At the last second, the hero Perseus saved the princess, however, while defeating his opponent, Cassiopeia and Cepheus were to turned stone by the head of Medusa. Poseidon then placed the king and queen in the sky and, as eternal punishment, Cassiopeia was condemned to circle the celestial pole forever, spending half the year upside down in the sky.

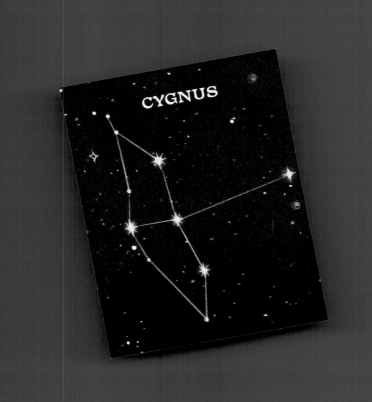

Identifying Features: It is known to some as the Northern Cross because of its five stars, Aljanah, Deneb, Fawaris, Sadr, and Albireo, that form its cross shape and run through the Milky Way.

Hemisphere: Northern. It's been seen June to December at latitudes +90° and -40°, but can be seen in the Southern Hemisphere low on the northern horizon in the winter.

Fun Fact: The brightest star in Cygnus is named Deneb, which sits at the tail of the constellation and whose name comes from an Arabic word for swan.

Nearby Constellations: Cepheus, Draco, Lacerta, Lyra, Pegasus, and Vulpecula.

Mythos: The swan is famous in Greek mythology because of the tale of Leda and the swan, who was really Zeus. Cygnus may also have gotten its name from the Chinese myth about the magpie bridge, Que Qiao (鹊桥). In the story, the lovers Niu Lang and Zhi Nu are separated by the Goddess of Heaven because Zhi Nu is a fairy and Niu Lang is human. When the goddess learns the two have married and had children in secret, she takes Zhi Nu and creates a river, the Milky Way, in the sky to separate the couple. Niu Lang goes to Heaven with his children so they can all be together, but the goddess doesn't relent. The story goes that once a year all the magpies in the world assemble to help the lovers be together by forming a bridge over the river, which is the constellation Cygnus.

Identifying Features: At a first glance, the seventh largest constellation in the sky might not appear to look like the winged horse it's named after, but that's because it is upside down. Flipped, you can make out the neck and head of a horse and two legs from the "Square of Pegasus," a major asterism made up of the following stars: Scheat, Alpheratz, Markab, and Algenib.

Hemisphere: Northern. It's best seen July to January at latitudes between +90° and -60°. It can be seen in the Southern Hemisphere August to December.

Fun Fact: The constellation is famous for hosting the first exoplanet (a planet found outside of the solar system) ever discovered around a yellow star, and it is also the home of the galaxy M15.

Nearby Constellations: Andromeda, Aquarius, Cygnus, Pisces, and Vulpecula.

Mythos: In Greek mythology, Pegasus is a white winged horse that sprang from the neck of Medusa when Perseus beheaded her. The name Pegasus comes from the Greek word *pegai*, which means "springs" or "waters." After his birth, Pegasus flew to Mount Helicon, where the nine Muses lived, and there he created the spring Hippocrene by striking the ground with his hooves. The name *Hippocrene* means "the horse's fountain," and it was believed that those who drank from the spring were blessed with the gift of poetry.

During your next outdoor hang with friends, see what constellations you recognize and can find. We know they'll be impressed by your knowledge of these constellations and the mythologies that shaped them.

Happy stargazing!

This book has been bound using handcraft methods and Smyth-sewn to ensure durability.

Designed by Justine Kelley.